Tobias Goes Seal Hunting

by Ole Hertz
Translated from the Danish by Tobi Tobias

Carolrhoda Books, Inc.   Minneapolis

This edition first published by Carolrhoda Books, Inc., 1984
Original edition first published by The Danish National Museum, Copenhagen, 1981
under the title TOBIAS PÅ SAELFANGST
Copyright © 1981 by Ole Hertz
English translation copyright © 1984 by Tobi Tobias
All rights reserved.
Published in agreement with International Children's Book Service, Copenhagen, Denmark
Manufactured in the United States of America

LIBRARY OF CONGRESS CATALOGING IN PUBLICATION DATA

Hertz, Ole.
  Tobias goes seal hunting.

  Translation of: Tobias på saelfangst.
  Summary: In Greenland, a boy and his father hunt seals
in their kayaks.
  [1. Sealing—Fiction. 2. Greenland—Fiction]
I. Tobias, Tobi. II. Title.
PZ7.H432463Tr   1984   [E]   83-26357
ISBN 0-87614-262-5 (lib. bdg.)

1 2 3 4 5 6 7 8 9 10 93 92 91 90 89 88 87 86 85 84

Tobias lives in a settlement in Greenland.
He is twelve years old.
He lives with his father and mother,
his big sister, Jensigne, and his little brother, Peter.

This is Tobias's house.
His father and mother built it themselves
from wooden boards.
There are only a few houses in Tobias's settlement,
and they are all like this one.
Tobias's father has a motorboat,
and a kayak, and a dogsled.
Tobias has a kayak too.

Today Tobias is going seal hunting with his father.
Tobias carries his kayak down to the motorboat,
and there it gets lashed tightly to the boat,
just like his father's kayak.

Tobias and his father sail for a long time
until they come to a fjord filled with ice.
There are usually seals here.
Tobias's father plans to sell the sealskins.
The family will eat the seal meat themselves.
Where Tobias lives, fishing and hunting
are the only ways people can earn money.

Now they shut off the motor and cast out the anchor.
From here they will sail on in their kayaks,
for in them they can silently steal up on the seals.

Now they paddle as quietly as they can,
farther into the icy fjord to look for seals.
They are careful not to splash with their paddles,
for the seals can be very shy.
Now and then they hold completely still
and listen to see if they can hear a seal breathing.

They sail for a long time without seeing any seals.
Once in a while Tobias's father
sticks his paddle down in the water
and knocks on it with a knife.
The sound may lure the seals.
They can be very curious.

Suddenly a seal pops up in front of them.

Tobias shoots at the seal,
but it is too far away,
so he misses it.
The seal dives down and swims away under the water.
The next time it comes up for air,
it hides itself behind an ice floe,
and they don't see it any more.

Fortunately, a little later Tobias's father shoots another seal
that is just as big as the first one they saw.
He ties the seal firmly to the side of his kayak
and fastens a little inflated bladder to it
so that it floats easily and isn't so heavy to tow.

Now they paddle back to the motorboat.
It has gotten too late for them to wait for other seals.
They must see about getting home.

Back at home, Jensigne has borrowed her father's binoculars.
Now she can see the motorboat coming, far away.
She calls her mother and her little brother,
and they all walk down to the shore.

Now the motorboat has come all the way in.
The seal must be taken ashore and carried up to the house.
It is hard to carry
because it is wet and slippery.

Tobias's mother flenses the seal.
Carefully she cuts the skin free from the body.
Later she will scrape, wash, and dry the skin
so that it can be sold.
Next she removes the meat that the family will eat.
She puts the rest in a bucket
for Jensigne to take to the others in the settlement.
It is always nice to get a gift of fresh meat.

Before long, delicious seal meat is cooking in the pot.
It will be good to eat something warm after such a long day.

A Greenlandic kayak is made of sealskin or painted canvas which is pulled over a wooden frame. Each tool has its special position on the kayak:

1 *The harpoon. After the seal has been shot, the harpoon is thrown so that the seal can be pulled to the kayak.*
2 *The harpoon line that ties the point of the harpoon to the bladder*
3 *The kayak chair. Most of the harpoon line lies coiled up on the kayak chair so that it can be tossed out quickly when the harpoon is thrown.*
4 *The hunting bladder is often made of an inflated seal skin, but it can also be just a plastic container. The hunting bladder prevents the seal from sinking when it is hit by the harpoon.*
5 *The kayak knife*
6 *A bag for gun and ammunition*
7 *The paddle*
8 *The kayak rudder which keeps the kayak on course*

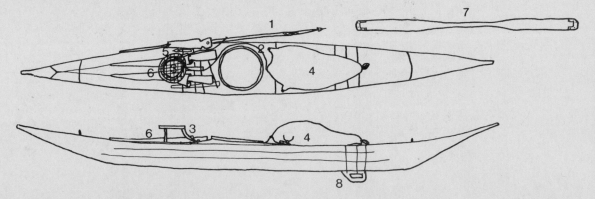